D1104981

Designing Products and Services That Customers Want

Management Master Series

William F. Christopher
Editor in Chief

Set 3: Customer Focus

Karl Albrecht
Delivering Customer Value: It's Everyone's Job

Robert King
Designing Products and Services That Customers Want

Wayne A. Little
Shared Expectations: Sustaining Customer Relationships

Gerald A. Michaelson
Building Bridges to Customers

Eberhard E. Scheuing
Creating Customers for Life

Ron Zemke
Service Recovery: Fixing Broken Customers

Designing Products and Services That Customers Want

Robert King

PRODUCTIVITY PRESS
Portland, Oregon

Management Master Series
William F. Christopher, Editor in Chief

Productivity Press
P.O. Box 13390
Portland, OR 97213-0390
United States of America
Telephone: 503-235-0600
Telefax: 503-235-0909
E-mail: staff@ppress.com

Book design by William Stanton
Cover illustration by Paul Zwolak
Graphics and composition by Rohani Design, Edmonds, Washington
Printed and bound by Data Reproductions Corporation in the United States of America

Library of Congress Cataloging-in-Publication Data

King, Bob, 1946–
Designing products and services that customers want / Bob King.
p. cm. -- (Management master series)
Includes bibliographical references and index.
ISBN 1-56327-145-1 (hardcover)
ISBN 1-56327-092-7 (paperback)
1. New products—Planning. 2. Quality function deployment 3. Customer services—Planning. 4. Consumers' preferences—Research. I. Title. II. Series.
TS170.K56 1995
658.5' 75--dc20 95-12447
CIP

00 99 98 97 96 95 10 9 8 7 6 5 4 3 2 1

—CONTENTS—

PUBLISHER'S MESSAGE

The *Management Master Series* was designed to discover and disseminate to you the world's best concepts, principles, and current practices in excellent management. We present this information in a concise and easy-to-use format to provide you with the tools and techniques you need to stay abreast of this rapidly accelerating world of ideas.

World class competitiveness requires managers today to be thoroughly informed about how and what other internationally successful managers are doing. What works? What doesn't? and Why?

Management is often considered a "neglected art." It is not possible to know how to manage before you are made a manager. But once you become a manager you are expected to know how to manage and to do it well, right from the start.

One result of this neglect in management training has been managers who rely on control rather than creativity. Certainly, managers in this century have shown a distinct neglect of workers as creative human beings. The idea that employees are an organization's most valuable asset is still very new. How managers can inspire and direct the creativity and intelligence of everyone involved in the work of an organization has only begun to emerge.

Perhaps if we consider management as a "science" the task of learning how to manage well will be easier. A scientist begins with an hypothesis and then runs experiments to observe whether the hypothesis is correct. Scientists depend on detailed notes about the experiment—the timing, the ingredients, the amounts—and carefully record all results as they test new hypotheses. Certain things come to be known by this method; for instance, that water always consists of one part oxygen and two parts hydrogen.

We as managers must learn from our experience and from the experience of others. The scientific approach provides a model for learning. Science begins with vision and desired outcomes, and achieves its purpose through observation, experiment, and analysis of precisely recorded results. And then what is newly discovered is shared so that each person's research will build on the work of others.

Our organizations, however, rarely provide the time for learning or experimentation. As a manager, you need information from those who have already experimented and learned and recorded their results. You need it in brief, clear, and detailed form so that you can apply it immediately.

It is our purpose to help you confront the difficult task of managing in these turbulent times. As the shape of leadership changes, the *Management Master Series* will continue to bring you the best learning available to support your own increasing artistry in the evolving science of management.

We at Productivity Press are grateful to William F. Christopher and our staff of editors who have searched out those masters with the knowledge, experience, and ability to write concisely and completely on excellence in management practice. We wish also to thank the individual volume authors; Diane Asay, project manager; Julie Zinkus, manuscript editor; Karen Jones, managing editor; Lisa Hoberg and Mary Junewick, editorial support; Bill Stanton, design and production management; Susan Swanson, production coordination; Rohani Design, graphics, page design, and composition.

Norman Bodek
Publisher

INTRODUCTION

This book is designed to introduce the reader to some of the key concepts and tools for improving customer satisfaction with products and services. The Kano three-arrow concept shows how to think about expected, one-dimensional, and exciting quality. Quality function deployment (QFD) shows how to prioritize customer demands, develop breakthroughs in product and service, and monitor implementation of new approaches. In QFD, the voice of the customer and voice of the engineer/specialist is geared to integrating the wants of the customer with existing organizational know-how. This helps organizations develop products and services that attract customers and grow market share.

1

THE KANO THREE-ARROW MODEL

WHAT IS CUSTOMER SERVICE QUALITY?

When you think of customer service, it is helpful to think of three categories of quality.

- *Expected quality,* which people take for granted until it is missing.
- *One-dimensional quality,* which makes people happy when they have it and unhappy when they don't have it.
- *Exciting quality,* which customers don't expect, but which brings excitement when you surprise them with it.

We can best understand these three kinds of customer service quality through the three-arrow model developed by Noriaki Kano, shown in Figure 1.

The vertical axis represents the level of customer satisfaction from low to high. The horizontal axis represents the extent to which customer requirements are met or not met.

The arrow that represents expected quality (a) shows that if customer needs are not met the customer is unhappy; if the needs are met it is not a big deal to the

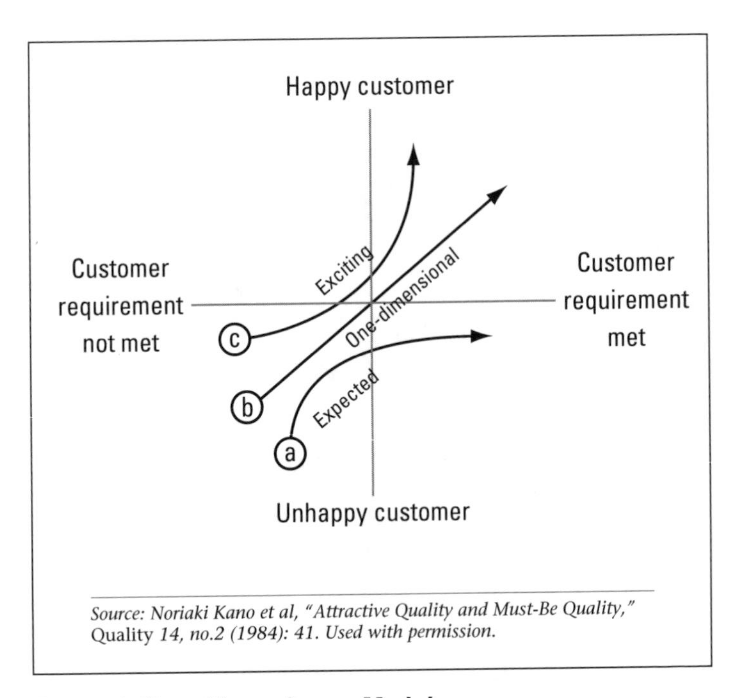

Figure 1. Kano Three-Arrow Model.

customer. An everyday example is paper in the restroom. If there is no paper, then it can be very unpleasant. If there is paper, then that is the way it should be—no big deal.

The arrow that represents one-dimensional service quality (b) shows that if you perform well the customer is satisfied; if you don't perform well the customer is dissatisfied. An example of this is waiting for service in a bank. If you get quick service, no waiting, or a short wait, you are happy. If you have to wait ten or more minutes for service, you are very unhappy.

The arrow that represents exciting service quality (c) shows that if you provide a service that makes a customer's life better, then you may excite and delight them. One service that has exciting quality is the ability to get the phone company to dial a number for you when you call information. If the service was free it would be really exciting.

ASSESSING CUSTOMER SERVICE QUALITY

Kano has developed a methodology to identify which aspects of customer service are expected, which are one-dimensional, and which are exciting.[1] But you can also assess these fairly easily yourself. Unsolicited complaints are most often about expected quality. One-dimensional quality items are most often identified by surveys in which customers rank wants from 1 to 5 or 1 to 9. The variation shows the more important one-dimensional quality items. The exciting quality items are the ones that the supplier develops, based on new insights and breakthroughs that make new modes of customer product or service possible.

Obtaining and clarifying the voice of the customer is an important task. Sources of customer input are market surveys, focus groups, warranty claims, and interviews. Regardless of the methods chosen, it is imperative to obtain the voice of the customer. Figure 2 is a cause-and-effect diagram that illustrates some of the considerations involved in capturing customer wants and needs.

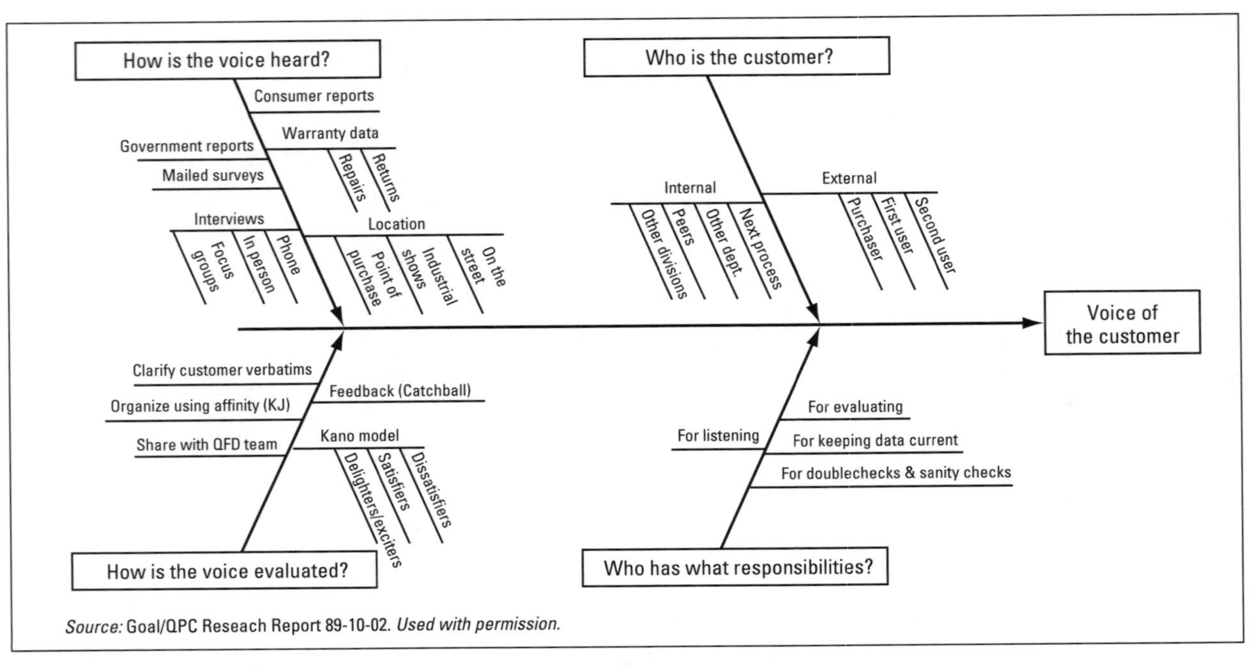

Source: Goal/QPC Reseach Report 89-10-02. *Used with permission.*

Figure 2. Voice of the Customer Cause-and-Effect Diagram

2

QUALITY FUNCTION DEPLOYMENT (QFD)

Customer satisfaction can be improved only when we hear and act on the voice of the customer. Using Kano's three-arrow quality as we actively listen to customers (through surveys, interviews, focus groups, complaints, and other means), we will identify needed improvements. Customer satisfaction is a moving target, always with opportunity for improvement.

Quality function deployment (QFD) is a helpful process for improving those areas most important to the customer. It is an important part of total quality management—a companywide process that involves all employees in all departments, every day, in improving quality, cost, yield, and delivery processes and systems.

QFD integrates the many different improvement technologies. It broadens employee involvement. It uses the 7 Quality Control Tools (See Appendix A), continuous improvement, and standardization. It also may instruct employees in value engineering, reliability engineering, and new concept selection.

QFD, like all aspects of total quality management, applies the scientific method to management. This method has been popularized as the Deming or Shewhart Plan, Do, Check, Act cycle (PDCA). The following is a list of steps for applying the PDCA cycle to QFD. The rest of this chapter describes these steps and provides tools to accomplish them.

Plan

1. PLAN **Define the project.**
   - What product or service should we improve or develop?
   - Define the scope of the project.
   - Assemble the right team.
2. PLAN **Identify the breakthrough areas.**
   - What do customers want?
   - What is needed from the engineering perspective?
3. PLAN **Define the breakthroughs.**
   - Generate ideas for improvement.
   - Select the best improvement concepts.

Do

4. DO **(Re) Design the new/improved product or service with breakthroughs.**
   - Design the breakthroughs into major systems (services, processes).
   - Design the breakthroughs into parts (service procedures).
   - Reduce the cost of the product or service.
   - Improve the reliability of the product or service.
5. DO **Assure successful follow-through in production and delivery.**
   - Assure implementation of supplier quality.

- Assure implementation of process reliability through fault tree analysis (FTA) and failure mode and effects analysis (FMEA).
- Assure implementation of process control and inspection control.

Check

6. CHECK **Gather customer feedback on changes.**
   - Collect complaints.
   - Conduct surveys.

Act

7. ACT **Analyze and plan for future projects.**
   - Monitor trends in sales and customer satisfaction.
   - Conduct competitive benchmarking.
   - Identify new QFD studies.

1. PLAN—DEFINE THE PROJECT

Step/decision	Tools
1. What product or service should we improve or develop?	• Customer Window (ARBOR)
2. Define the scope of the project.	• Purpose hierarchy • TRIZ project questionnaire (Ideation)*
3. Assemble the right team.	• Systems questions

* For more information about TRIZ, see the glossary.

1. What Product or Service Should We Improve or Develop?

Often we are out of alignment with what customers want. We make good products that they don't want and make poor products that they do want. The first step in project definition is to align what we are doing with what customers want.

A helpful tool in doing this is the Customer Window™ developed by ARBOR.[2] The Customer Window provides a way to analyze various customer voices simultaneously.

By looking at the priorities of customers, we can focus our efforts on improving both satisfaction and efficiency. In the Customer Window in Figure 3, Box A is an area where customers are not getting what they want. Box B is an area of success the company needs to maintain. Boxes C and D are potential areas for cost cutting. In the case of Box C, we need to communicate with the customers to decide if we are overspending. Box D probably represents a double opportunity for savings. If we are not providing good programs or services, we may be wasting money, and if the customers do not want what we are providing, we may be wasting money.

Figure 3 illustrates a hospital's corporate health program. The example indicates a need to look at items 4 and 7 to determine why they are not satisfactory and to make improvements. The successful areas need to be maintained. Item 5, the use of illness prevention literature, needs to be examined. The long-term financial benefit to the corporation, in terms of avoidance of medical costs, is considerable. It needs to be determined what alternative approaches could be used if the literature is not helping. In the case of the weight loss program, employees tend not to want to attend sessions together, and the company

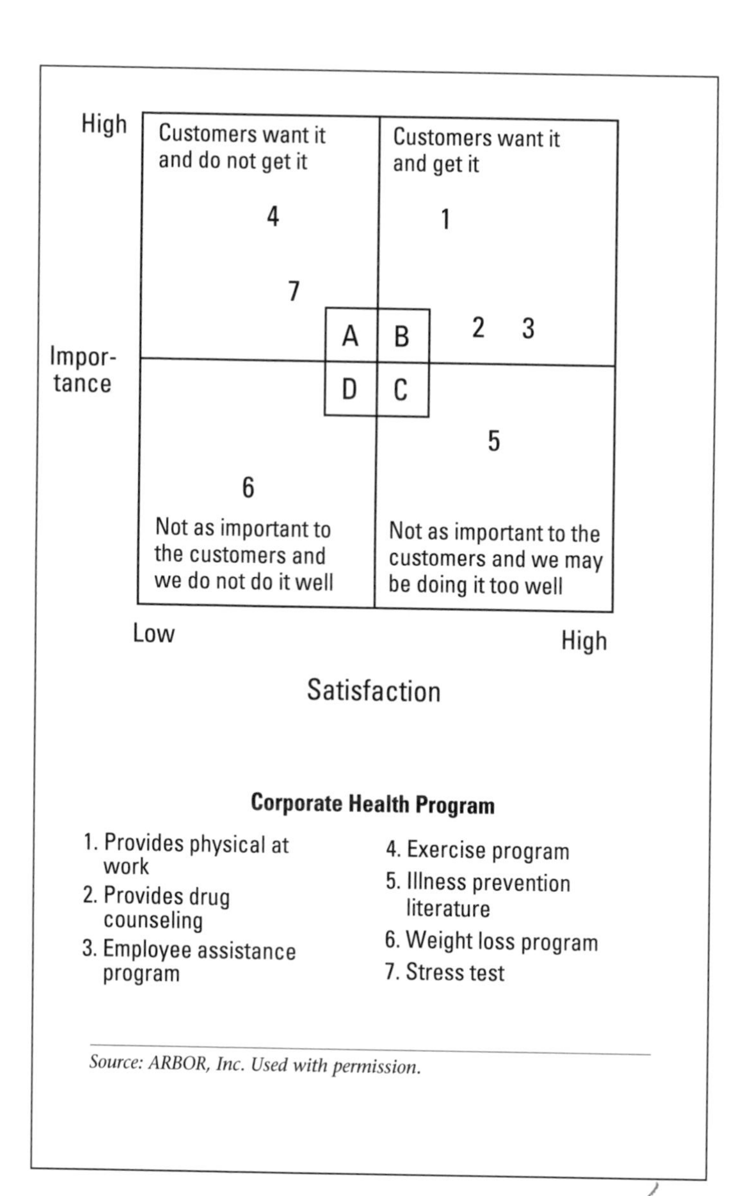

Figure 3. Customer Window Model for Sorting Customer Demands

is also competing against many fine programs in the community. Maybe this one gets dropped.

Customer focus and related tools of quality management are not quick cure-alls. We cannot use them exclusively. We must look at them in conjunction with an analysis of critical processes. But they do provide a helpful way of focusing on key issues that affect customer satisfaction and costs. And as more and more employees get involved in continuous improvement, customer focus is one way to steer efforts toward the vital few improvements that will make the most difference.

2. Define the Scope of the Project

Often we get into trouble because we define projects too broadly or too narrowly. One way to solve this problem is to generate an array of possible projects (from narrow to broad) and then pick the one most appropriate to work on at this time.

The *purpose hierarchy,* popularized by Gerald Nadler, is one way to do this.[3] Suppose we are considering a project on TQM and creativity. We can use the following purpose hierarchy to structure a brainstorming session to define the project in the following manner:

1. Identify the purposes.
2. Group the purposes from simple to complex, or short-term to long-term.
3. Identify the right level of the purpose hierarchy to work on.

The purpose hierarchy works best if you do the following:

- Write each separate purpose on a different Post-it.™

- Find the easiest purpose and label it S for simple. Code all the other Post-its:
  - *S* for simple, *S-M* for simple to medium,
  - *M* for medium, *M-C* for medium to complex,
  - *C* for complex, and *X-C* for extra complex.

A sample purpose hierarchy on the issue of integrating TQM and creativity is shown in Figure 4.

The team working on this project had already studied TQM tools and systems so the next project was to study innovation tools and systems.

Integrating TQM and innovation:

- (S) Study TQM tools and systems.
- (S) Study innovation tools and systems.
- (M) Replace TQM tools with superior innovation tools where appropriate.
- (M) Include innovation processes in TQM system where there are voids.
- (C) Develop fully integrated TQM/innovation system.

Figure 4. Sample Purpose Hierarchy

3. Assemble the Right Team

Another important step in the QFD process is to assemble the right team. Some possible considerations for team members are listed here:

- Members should represent all the functions necessary for successful implementation, for example:

  - Design
  - Manufacturing
  - Reliability
  - Purchasing
- Decision makers need to participate in, or at least support, the team.
- The team needs to be clear about how to get customer input:
  - Can marketing help?
  - Will engineers form focus groups?
- The team needs to think through the design process and determine who else needs to be involved for success. Not everyone needs to be physically present all the time. Some experts may only review documents and critique material.
- The core team should consist of four to eight people with specialists brought in as needed for specific areas.

2. PLAN—IDENTIFY THE BREAKTHROUGH AREAS

Step/decision	Tools
1. What do customers want?	• Voice of the customer table (Figure 5) • Generation of quality characteristics or metrics (Figure 6) • House of Quality (Figure 7)
2. What is needed from the engineering perspective?	• Tree diagram (Figure 8)

What Do Customers Want?

A voice of the customer table[4] allows you to gather ideas about customer wants and needs. Observing customers' use of the product or service can be helpful. Observations can be listed in a voice of the customer table, as shown in Figure 5.

Name of product/service ____________________
When it is used ____________________
Where it is used ____________________
Why it is used ____________________
How it is used ____________________
What it is...how the user describes it ____________________
What problems the customer had? ____________________
What does the customer like? ____________________
What does the customer dislike? ____________________

Figure 5. Voice of the Customer Table

Generation of quality characteristics or metrics categorizes information from the voice of the customer table into customer wants and metrics (also known as substitute quality characteristics). These wants and metrics identify to what extent the want is met.

Figure 6 summarizes customer information about getting a haircut. Customer words in the first column are translated into customer wants in the second column and metrics, or substitute quality characteristics, in the third column. The reason for the translation is that customers' words are not usually precise enough to be uniformly understood by engineers. Column 2 tries to give some precision to what customers want without changing the meaning. Column 3 identifies what can be measured or controlled to assure that customer wants are met.

Customer words	Customer wants: value statements	Metrics: How success is measured
Not too short	Looks like I expected it to look	Amount of hair cut is okay on first look in mirror
Doesn't look " just cut"	People like the way I look	Number of favorable comments
		Number of unfavor-able comments
Doesn't take too long	The barber/stylist is efficient	Time to cut hair
I don't have to wait long for a barber	I get seated quickly	Time to wait for a chair
Barber friendly, but not too nosy	I can talk or sleep during haircut	My level of comfort with barber-chair chit chat
People notice my haircut	My haircut is attractive	Number of positive comments I get on my haircut

Figure 6. Generation of Quality Characteristics or Metrics: Haircut/Styling Example

This is all very interesting and useful information. But the real advantage of QFD comes when you prioritize the metrics and do better than the competition. The way that the metrics are prioritized is through a matrix, as shown in Figure 7. This matrix is sometimes referred to as *the House of Quality.*

The number-one priority is the number of favorable comments. The close second is the length of time needed to cut hair.

The simple QFD matrix[5] is filled out in this way:

- List customer needs (as translated from customer words) on the left side of the chart.

Prioritizing the metrics

Legend: ⦿ Primary; ○ Secondary; △ Possible

Customer wants	Amount of hair cut is okay with first look in mirror	Number of favorable comments	Length of time to cut hair	Time to wait for a chair	My level of comfort with barber-chair chit chat	**Level of importance**
Looks like I expected it to look	⦿ 45	○ 15				5
People say they like it		⦿ 45				5
Barber/stylist is efficient			⦿ 36			4
I get seated quickly			○ 12	⦿ 36		4
I like the conversation with the barber/stylist			○ 9		⦿ 27	3
Total	45	60	57	36	27	

Figure 7. Prioritization Matrix (simple House of Quality)

- List metrics, or substitute quality characteristics, across the top.
- Down the right side, list the average rating of each customer demand on a scale from 1 to 5 (with 5 being high and 1 being low).

The middle of the matrix is filled with symbols that indicate the relationship between the two items:

- A strong relationship is a double circle ⦿ with a value of 9.
- A moderate relationship is a single circle ○ with a value of 3.

- A possible relationship is a triangle △ with a value of 1.

The value of the relationship is multiplied by the level of its importance, giving weighted value for that box. For example, in Figure 7, in the box in the upper left corner, a strong relationship (⦿ = 9) is multiplied by the highest level of importance—giving a total of 45. This procedure (weight of relationship times weight of importance) is done for all the boxes in the matrix. Then each column is added, giving the total weight for each metric.

The item with the highest rating is the number of favorable comments, followed by length of time to cut hair. This means that if the barber/stylist figures out how to give a haircut that looks good in better time than the competition, he or she will have an advantage in the marketplace.

The customer does not know everything. One of the most humorous stories of the 1980s concerns a U.S. car company that conducted focus groups to find out what people wanted from their catalytic converters. Customers didn't want them to break and they didn't want them to smell. These are important data points, but are not sufficient to design a catalytic converter. Much more is needed, for example, its purpose and how it works. This is why we need both the voice of the customer and voice of the engineer/specialist.

2. What Is Needed from the Engineering Perspective?

The voice of the customer can be explained in a *tree diagram* as shown in Figure 8.

QFD comes out best if the tree is balanced with a similar number of detailed items. For example, in Figure 8, each of the three primary items has two subitems at the level of detail.

- Customer wants
  - Appearance
    - Looks like I expected it to look
    - People say they like it
  - Timeliness
    - Barber/stylist is efficient
    - I get seated quickly
  - Pleasant atmosphere
    - I like the conversation with the barber/stylist
    - There is a good selection of reading material

Figure 8. Haircut Tree Diagram—Voice of the Customer

- Role of the barber/stylist
  - Provides an attractive haicut
    - Knows how to give various haircuts
    - Knows what the styles are
  - Provides the haircut the way customers want
    - Asks person what he or she wants
    - Suggests what might look good
  - Keeps the area clean
    - Cleans up after each haircut
    - Uses a clean comb and equipment for each haircut

Figure 9. Haircut Function Tree—Voice of the Engineer

⊙ Primary
○ Secondary
△ Possible

Voice of the engineer/specialist \ Voice of the customer	Looks like I expected it to look	People say they like it	Barber/stylist is efficient	I get seated quickly	I like the conver-sation with the barber/stylist	There's a good selection of reading material
Knows how to give various haircuts	○	⊙	○	○		
Knows what the popular styles are	○	⊙	○			
Asks person what he/she wants	⊙				○	
Suggests what might look good	○	⊙	△		○	
Cleans up after each haircut						
Uses clean comb and equipment						

Figure 10. Match-up Matrix

The voice of the engineer is best conveyed in a function tree, as shown in Figure 9. Each function statement should have a verb and an object.

These two tree diagrams can then be combined into a match-up matrix, as shown in Figure 10.

What we can learn from this depends on the gaps. No voice-of-the-customer items correlate with the voice-of-the-engineer/specialist, "clean up after each haircut." Either the customer doesn't care, or more likely, hasn't been asked. Also there is not much correlation between the specialist item, "ask the person what he or she wants," and the voice of the customer.

If you can match the customer voice with the engineer/specialist voice completely, you are likely to go a long way with customer satisfaction.

3. PLAN—DEFINE THE BREAKTHROUGHS

Step/decision	Tools
1. Generate ideas for improvement.	• 7 Creativity Tool Boxes • TRIZ
2. Select the best improvement concepts.	• New concept selection

1. Generate Ideas for Improvement

In the past, creativity has been limited in organizations. One reason is that few creativity authors focus on step-by-step approaches to how to do creative work.[7] Another reason is that in a departmentalized environment many think that only marketing and development employees should be taught creativity. Since training costs money, we put the effort where we think it will do the most good.

A third reason creativity is limited is that many organizations have a climate of fear, with many habits that constrain or reduce creativity. In an organization still tied to the departmentalized Taylor model, a few people have the job of improvement, while most others have the job of carrying it out or inspecting it. In that scenario, creativity training only makes sense for the few involved in improvement.

Total quality management (TQM) and continuous quality improvement (CQI) over the last 30 years have changed that. Today, many organizations have adopted the idea that all employees in all departments should be involved in improving and maintaining quality, cost, yield, processes, and systems in order to give customers products and service that are most economical and best qualified. In this scheme of things, all employees should have the benefit of creativity tools to increase idea generation and the successful implementation of the best ideas.

Two methods for generating ideas for improvement are the 7 Creativity Tools and TRIZ. These creativity tools are discussed here and in Appendix C.

BRAINSTORMING

One of the best known methods for collecting ideas is *brainstorming*. This method, which randomly generates and lists ideas, is an important tool for both problem solving and planning. Brainstorming is often divided into unstructured and structured activities. Using unstructured brainstorming, group members simply give ideas as they come to mind. It tends to create a more relaxed atmosphere, but also risks domination by the most vocal members.

Using structured brainstorming, every person in a group must give an idea as their turn arises in the rotation or pass until the next round. It often forces even shy people to participate, but can also create a certain amount of pressure to contribute.

Several "rules of the road" have been found helpful for brainstorming:

- **Do not criticize any idea**. There should be no negative comments or criticism of people's suggestions. Once we begin to evaluate ideas, we leave the generation of ideas behind. The recorder must write each idea in the words of the speaker on an overhead or flipchart.

- **Be an eager listener.** Take what you hear as the basis for further development. Focus on what is positive in each idea. One approach to brainstorming requires participants to say something positive about a past idea before they continue with their own idea.
- **Say any ideas that come to mind.** Don't hold back. We tend to hold back from speaking until we have the perfect idea. One should share even parts of ideas.
- **Produce many ideas.** Quantity, not quality. This produces spontaneity. Ideas should be complete sentences, but confined to essentials.

BRAINWRITING

Another method for collecting ideas is brainwriting—a kind of structured brainstorming in which the participants themselves write down all the ideas. Method 6-3-5 was invented in the late 1960s by Bernd Rohrbach in Germany. This method involves six people brainstorming ideas and writing them down. Each person takes turns brainstorming and writing three ideas, then passes the paper to the person on the right five more times until there are 6 sheets with 18 ideas for a total of 108 ideas (takes about half an hour), as shown in Figure 11.

The 6-3-5 structured brainwriting method, of course, requires agreement on the topic. If some people get new ideas from one or more previous ideas, they can list those ideas as well. Select one person to be a timekeeper/facilitator. The average time for each box is five minutes, with the first participants taking a little less time and the later participants taking a little longer so they can read previous entries.

The advantages to the 6-3-5 method of idea generation are:

Method 6-3-5

	1	2	3
1	Look at fringe technologies.	Get a psychiatrist on the team.	Find explorers.
2	Address spirituality.	Look at field theories (magnetic, heat, energies, etc).	Look at original theories of cognition: " How do we think?" Dr. Maturana Dr. Verala
3	Integrate creativity knowledge.	What are types (M/B) and influences on creative process.	Study creative people as models.
4	Interview leaders in field.	Try current radical approaches to under-stand benefits of teaching people to be more creative.	Develop matrix of creative people. Use TQM tools to prioritze.
5	Dare to be different. Take risks. How do we create a safe environment for others?	Reward creativity in a creative way.	Demand creative leaders. How do you recognize creativity in the interviewing process?
6	How does the organization convince people that taking risks is *really* "all right"?	Don't force people to believe that only "creat-ive" solutions are valued by the orginization. Don't make "lack of creativity" a new orginizational fear/label.	What kinds of support do middle managers need to allow "risking" practices to proceed?

Figure 11. Brainwriting question: What does the creativity/ innovation team need to consider?

- Writing rather than speaking the ideas prevents criticism.
- Ideas come on to the paper originally; the recorder changes nothing.
- The 6-3-5 group can have more people—20 or more—brainwriting, as long as there are more

people to collect the ideas and more space to post them. This works well for big conferences.

Some disadvantages to using the 6-3-5 method are:

- Space for writing is limited. Technical problems, for example, may require more space for drawings.
- The tendency is to go with catch-phrases, so ideas may not be fully clear to all.
- It is difficult for the moderator to provide coaching or help.
- The method may be less interesting, because a well-constructed brainstorm can be a lot of fun.
- The large proliferation of ideas may result in duplication.

In addition to brainstorming and brainwriting (using the 6-3-5 method), some additional creativity tools for generating ideas are:[8]

- *Creative brainstorming,* in which we use a foolish topic like getting employees to wear cardboard noses to break out of current thinking and add new ideas to the brainstorming.
- *Picture and word simulation,* in which we use words and pictures to stimulate the generation of additional ideas.
- *Analogies,* which provide a way to build new solutions from parts of past solutions. The TRIZ system taps the analogies of over one million inventions to generate new ideas.
- *Morphological charts,* which generate the parameters of the solution and attempt to capture all possible solutions to the problem.

2. Select the Best Improvement Concepts

To select the best improvement concepts a useful tool is new concept selection, made popular by the late Stuart Pugh of Scotland. There are many approaches to getting breakthroughs in QFD.[9] For the haircut example a simple new concept selection chart may help, as illustrated in Figure 12.

		New Concept					
		Use basic models	Cut a lot at once	Get accurate explanation of what person wants	Have pictures clients use to choose the look they want	Datum	Have pictures of various looks available during client waiting time
Criteria	Favorable comments	+	+	++	++	Haircuts, Inc.	++
	Quick haircut	++	++	+	+		++
	TOTAL	3+	3+	3+	3+		4+

Figure 12. New Concept Selection Chart

The new-concept selection process lists the criteria on the left and the new concepts across the top. The datum column contains the name of an organization that acts as the current standard. In the chart, a plus (+) indicates positive, two pluses (++) indicate very positive, a minus (−) indicates negative, and two minuses (−−) indicate very negative. The total number of pluses in the columns suggest which concepts meet more of the criteria.

Figure 12, for example, shows that the salon could provide a book with pictures of sample haircuts. People could then review them while they are waiting, putting

Post-its on the pages they like. The barber/stylist would then get a clear picture of what the haircut should look like. Barbers/stylists could be trained on how to create all the various looks quickly.

All of the previous stages have been geared to defining the PLAN part of the PDCA cycle. We must move to the Do—the design and production of products and services that will delight customers. We must dot the i's and cross the t's in this part of the process or we will fall short of delight.

4. DO—(RE)DESIGN THE NEW/IMPROVED PRODUCT OR SERVICE WITH BREAKTHROUGHS

Step/decision	Tools
1. Design breakthroughs into major systems (services, processes).	• Process and new technology matrix
2. Design breakthroughs into parts (service procedures).	• Parts and customer wants matrix
3. Reduce the cost of the product or service.	• Value engineering
4. Improve the reliability of the product or service.	• Fault tree analysis (FTA) • Failure mode and effects analysis (FMEA)

1. Design Breakthroughs into Major Systems (Services, Processes)

Breakthroughs usually require the application of new technology. They can be found by looking at the major aspects of the product (service) in relation to new technology, as illustrated in Figure 13.

New technologies	Haircutting processes	Hair styling processes	Processes for a pleasent environment	Processes to describe cut	Processes to evaluate job during cut
Video technology	⊙				⊙
Computer technology				⊙	
Earphones with music chair			⊙		

Processes

⊙ Primary
○ Secondary
△ Possible

Figure 13. New Technology Matrix

This chart helps explain which new technologies provide improvement opportunities. Across the top are processes related to various functions. The vertical column lists new technologies. Video technology shows some promise with two double circles.

2. Design Breakthroughs into Parts (Service Procedures)

One method for designing breakthroughs into parts is to compare product or service details with customer demands.

People like me, who wear glasses for distance, can't easily evaluate a haircut until it is done. Easily available computer technology would make it possible for these people to watch the haircut on a closely placed, ceiling-mounted video monitor.

A remote zoom control could allow a full view or a close-up of what is going on. A matrix for comparing products or service details with customer demands is shown in Figure 14.

Priortized customer wants

⦿ Primary
○ Secondary
△ Possible

Product service details		Number of times haircut is right	Number of favorable comments	Length of time to cut hair
Video	Remote focus	⦿	○	○
Video	Adjust mount of monitor	⦿		○
Discussion about how to cut hair		⦿	○	○
Feedback on how customer likes haircut		⦿	○	○

Figure 14. Designing Breakthroughs into Parts

3. Reduce the Cost of the Product or Service

Value engineering provides an easy graphical way to see where to make improvements. Elements of a product or service are plotted on a graph, as shown in Figure 15. The vertical axis shows the value of the element or component. The horizontal axis shows the cost of the element or component. In the example, the discussion about how the haircut is going has high value and low cost. So, it is plotted above the diagonal line, which is where relative value and cost are equal. Installing a video camera and monitor so the customer can always see the back of his or her head has a high cost compared to the value.

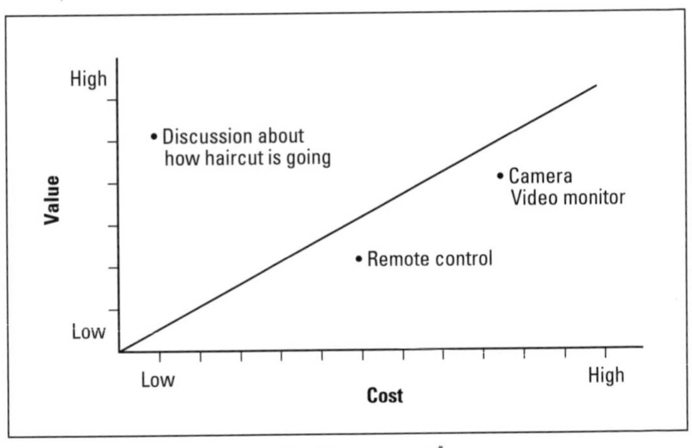

Figure 15. Value Analysis Chart

The value engineering process is a way of identifying items whose relative cost is high compared to their value. These then become the target for cost reduction. It is important not to reduce quality while reducing cost.

We especially want to look at the cost of the video camera and monitor. Although the value is high, the cost is proportionally higher. Is there a way to take the cost out of this? Is there another way to provide a greater quality with less cost?

4. Improve the Reliability of the Product or Service

The fault tree analysis (FTA) organizes the things that can go wrong from major to minor categories, as shown in Figure 16.

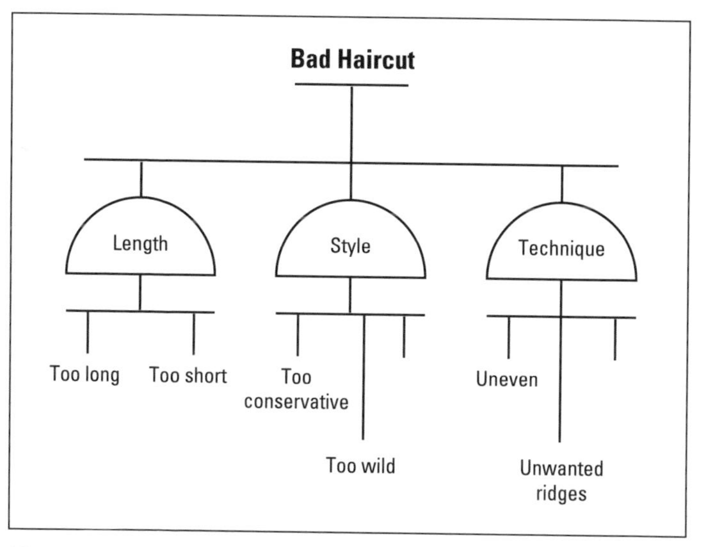

Figure 16. Fault Tree Analysis

The failure mode and effects analysis (FMEA) is a table that prioritizes faults based on frequency of detection and severity. These are ranked on a scale of 1 to 10, with 10 as a serious problem and 1 as a minor problem, as shown in Figure 17.

The risk priority is the product of *severity x frequency x detectability*

Example: Too short *4 x 3 x 2 = 24*
Choppy *6 x 3 x 3 = 54*

This risk priority suggests working on *choppy* first. The creativity tools mentioned earlier, especially the morphological chart, are helpful for coming up with ideas to reduce these problems.[10]

	Severity	Frequency	Detectability	Risk priority
Too short	4	3	2	24
Choppy	6	3	3	54

Figure 17. Failure Mode and Effects Analysis Table 

5. DO—ASSURE SUCCESSFUL FOLLOW-THROUGH IN PRODUCTION AND DELIVERY

Step/decision	**Tools**
1. Assure implementation of supplier quality.	• Assign an engineer or specialist from the customer and supplier organizations
2. Assure process reliability.	• Fault tree analysis • Failure mode and effects analysis
3. Assure implementation of process control and inspection control.	• Quality Control Process chart • Feedback metrics

1. Assure Implementation of Supplier Quality

Supplier quality depends on the supplier quality process. Clearly you can't fix defects from your suppliers and succeed in delighting your customers. You must move toward working with suppliers who provide right quality 100 percent of the time and deliver when you want it 100 percent of the time.

This can be accomplished by assigning an engineer or specialist from the customer and from the supplier organization. Each lists all the obstacles the other creates to getting 100 percent quality on time, all the time. They then prioritize and eliminate these, one at a time, until they reach 100 percent.

2. Assure Process Reliability

Process reliability, like product reliability, uses fault tree analysis (FTA) and failure mode and effects analysis (FMEA).

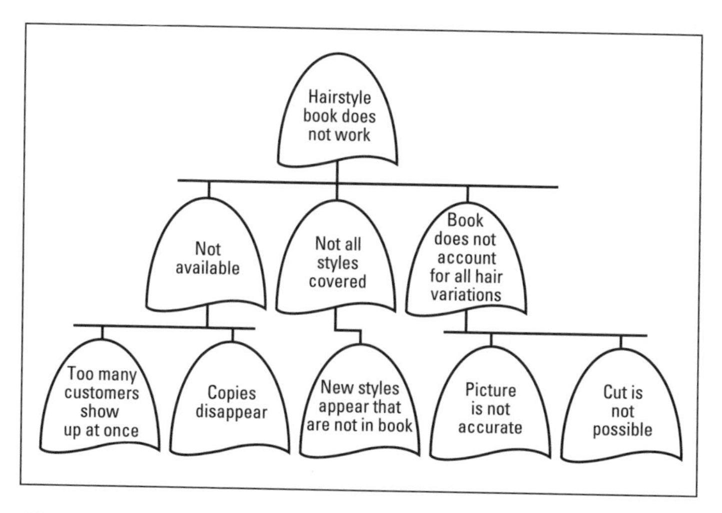

Figure 18. Process Fault Tree Analysis

FAULT TREE ANALYSIS

Fault tree analysis maps the relationships between each potential cause of failure. Process fault tree analysis maps out what can go wrong by using an upside-down tree diagram to show general categories of mistakes branching out to possible individual mistakes (see Figure 18).

FAILURE MODE AND EFFECTS ANALYSIS

The number of problems that can occur is limited only by the capacity of minds to conceive them or accidents to produce them. Failure mode and effects analysis (FMEA) helps prioritize the failures to work on. So the mistake "style does not appear in book" is slightly more important than "picture does not look like reality," as described in Figure 19.

	Severity	Frequency	Detectability	Risk priority II
Style does not appear in book	2	3	5	30
Picture does not look like reality	3	3	3	27

Figure 19. Failure Mode and Effects Analysis

3. Assure the Implementation of Process Control and Inspection Control

The final part of QFD is making sure it happens. The quality control process chart in Figure 20 helps monitor the effectiveness of the new approach.

PROCESS	MEASURE	WHO	TARGET	RESULTS
Person gets explanation sheet	Percent who get sheet	Receptionist	100%	
Person reviews book	Percent who need book	Customer	50%	
Person picks look	Percent who pick look	Customer	25%	
Barber cuts	Length of cut vs. standard feedback sheet	Barber	25% Reduction	

Figure 20. Quality Control Process Chart

There are several steps to improvement.

1. List the steps in the process. This can be done in a flowchart or simply by listing them.[11]
2. Identify the metric—what will be measured. List by whom and the target. This makes it possible to get real data. As Roger Milliken of Milliken Company said, "If you are not measuring, you are just practicing."[12]

The concept of customer focus is all about measuring how well we are satisfying customer expectations and

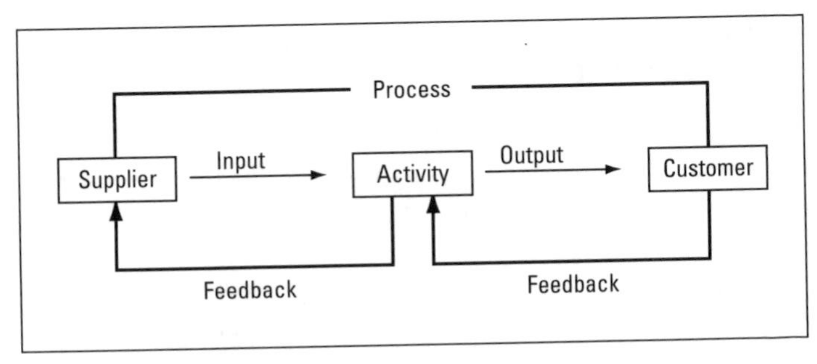

Figure 21. Feedback Metrics for Process Improvement

how much we are improving. The total quality effort is based on process improvement, as shown in Figure 21.

To perform an activity properly the input must be correct. That is why the best quality companies in the U.S. are reducing the number of suppliers, based on the quality of their input and their processes.

The voice of the customer provides data on the quality of the product or service. This data provides the basis for knowing what to improve and whether we are in fact improving.

6. CHECK—GATHER CUSTOMER FEEDBACK ON CHANGES

You need to see if you have delighted the customer. A classic example is the new design from Komatsu after they broke off from the joint venture with International Harvester in 1982. A portion of their QFD chart showed "operator comfort" and "not expensive to operate" as key customer needs. After designing and producing the new models, they saw positive responses, as indicated in Figure 22.

Both surveys and complaints are good ways to gather customer feedback. As we mentioned earlier in the section on Kano's three-arrow diagram, complaints are a

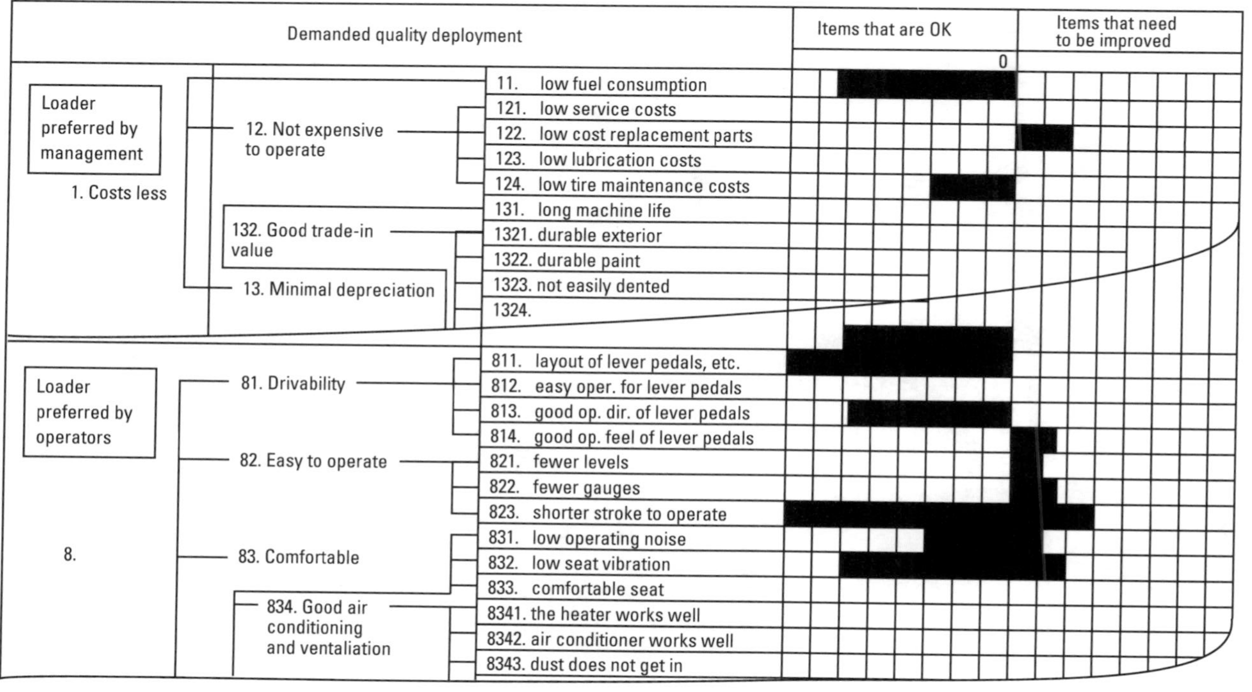

Figure 22. Demanded Quality Deployment Chart (Komatsu MEC Case)

helpful key to what customers consider expected quality. Since perhaps only 10 percent of customers ever complain, it is important to reassure those that do for the learning potential.

Processes must be in place to ensure that complaints are dealt with in a timely way, and to ensure that all the complaints are gathered and analyzed to see the frequency and areas of problems.

7. ACT—ANALYZE AND PLAN FOR FUTURE PROJECTS

Where do we go next?

Benchmarking

Benchmarking provides some help. It was popularized by the work of Robert Camp at Xerox.[13] It provides a way of gathering improvement ideas. It can't be seen as a method of copying because one is always behind the one they copy. Rather, benchmarking is a way of stimulating improvement ideas.

Benchmarking is often enhanced by doing it outside your industry. Federal Express may be particularly good at tracking orders. L.L. Bean may be good at customer service, and so on.

Another important point of benchmarking is to be open to unexpected discoveries. It is important not to define the benchmarking activity so narrowly that you miss the alternative ways people do things.

In benchmarking, one identifies the gap between processes from one company to another. This gap becomes a stimulus for change. Additional QFD charts may help define those changes.

Monitoring Trends

Monitor trends in sales and customer satisfaction. QFD should improve market share. If market share did not go up, QFD was unsuccessful. As growth in market share slows, it is time to practice more QFD.

CONCLUSION

Customer satisfaction is rooted in knowing what the customer wants (voice of the customer). The important thing in customer satisfaction is to prioritize what customers want and figure out how to achieve it better than the competition (QFD). It is important in customer service to satisfy customers' expected quality, one-dimensional quality, and exciting quality (Kano three-arrow quality).

The recommended reading list at the end of this book provides the opportunity for additional study on customer service. Good luck in satisfying your customers and building a successful business.

APPENDIX A
THE 7 QUALITY CONTROL TOOLS

CAUSE-AND-EFFECT DIAGRAM

The cause-and-effect diagram is also called the fishbone chart because of its appearance, and the Ishikawa chart after the man who popularized its use in Japan. It provides a convenient way to portray the primary and secondary causes of a particular effect. This aids in selecting the right problem to work on.

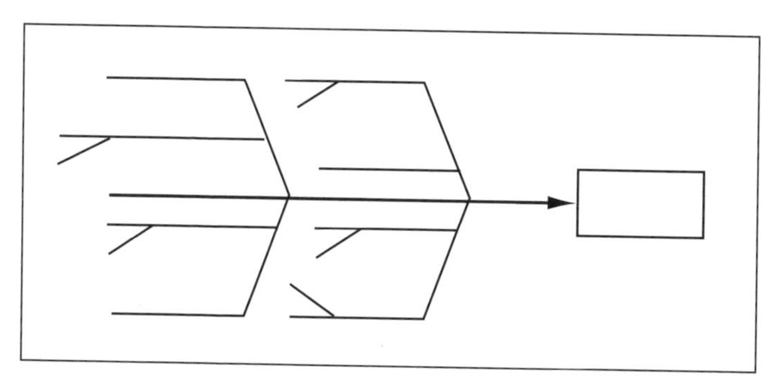

Cause-and-Effect Diagram

RUN CHART

The run chart shows the history of variation; whether a certain variable is going up or down. It is important to indicate on each run chart what is good or bad. This tool is beneficial in the beginning of a change process. It helps identify where the problem is and what the baseline is. It is also important at the end of the change process to measure what progress was made.

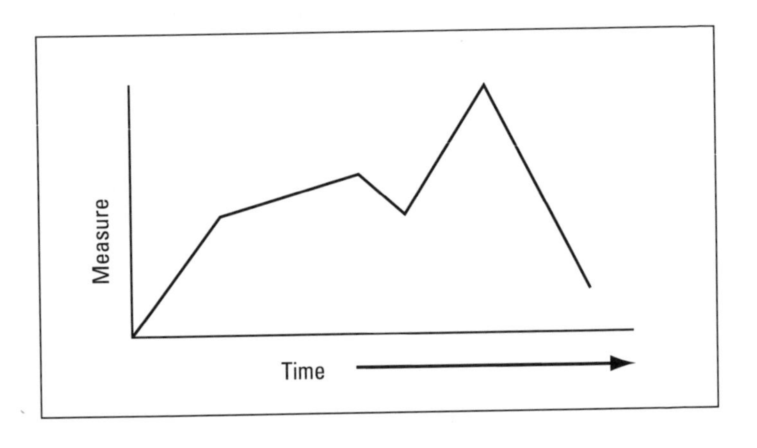

Run Chart

SCATTER DIAGRAM

The scatter diagram shows the relationship between two variables. Sometimes, as the chart below shows, one variable goes up as another variable goes up. Other times, one variable goes down as another variable goes up. The closer the points are to the center line of the cluster, the stronger the correlation. If the dots are widely scattered, there is little or no correlation.

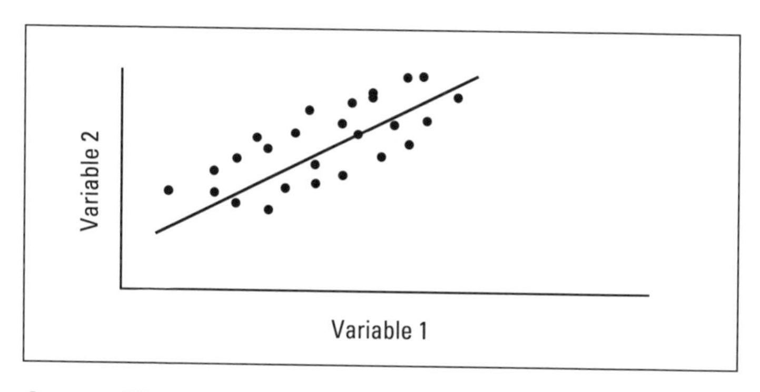

Scatter Diagram

FLOWCHART

Flowcharts identify the actual and ideal paths that any product or service follows in order to identify variations. The circle symbol ○ identifies the beginning or end of the process. The box □ identifies an activity and the diamond ◇ identifies a decision.

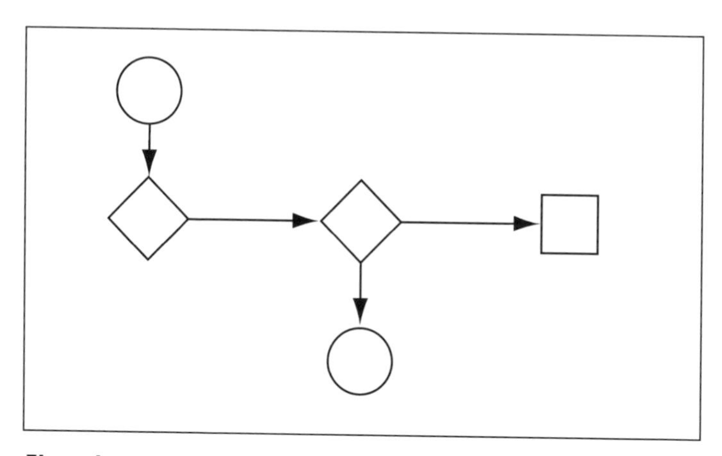

Flowchart

PARETO CHART

The Pareto chart identifies the quantity of an item. The bars are arranged in decreasing size from left to right. This makes the most frequent item quite visible. The last column is reserved for other.

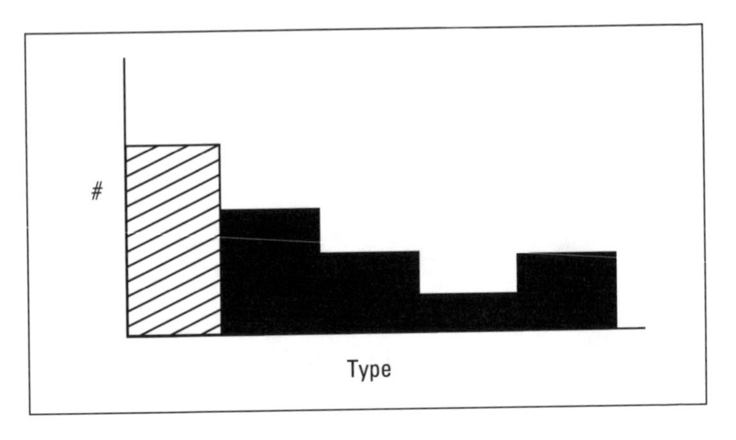

Pareto Chart

HISTOGRAM

The histogram shows the distribution of variation. The histogram below shows a single variation, well-balanced in the traditional bell-shaped curve. One benefit of the histogram is that it shows when the distribution is skewed to one side or another. Another benefit is to show the width of the variation. If it is wider than the standard, then it is impossible to meet the standard.

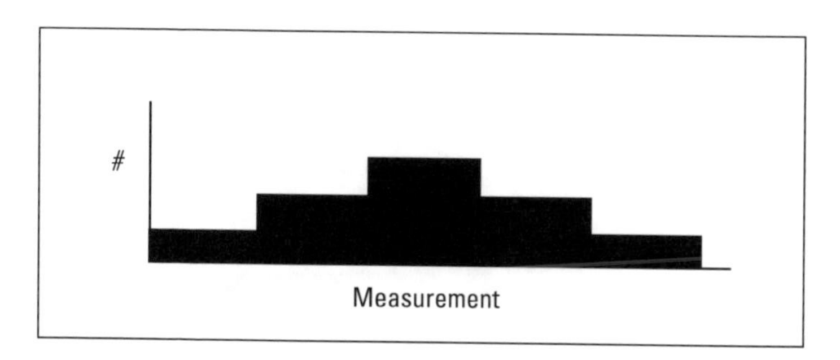

Histogram

CONTROL CHART

The control chart is a line chart with the benefit of statistically calculated control limits. These limits show what variation is expected in a system in control. If an item goes beyond the upper or lower control limit (UCL and LCL), then it can be recognized as a special cause. In continuous improvement activity it is recommended to eliminate special causes first and then common causes.

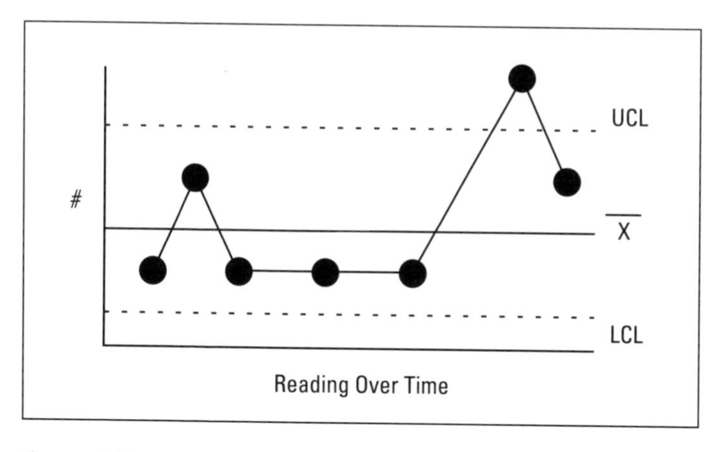

Control Chart

APPENDIX B
THE 7 MANAGEMENT AND PLANNING TOOLS

AFFINITY DIAGRAM

The affinity diagram gathers large amounts of language data (ideas, opinions, issues, and so on), organizes the data into groupings based on the natural relationship between each item, and defines groups of items. It is largely a creative, rather than a logical, process.

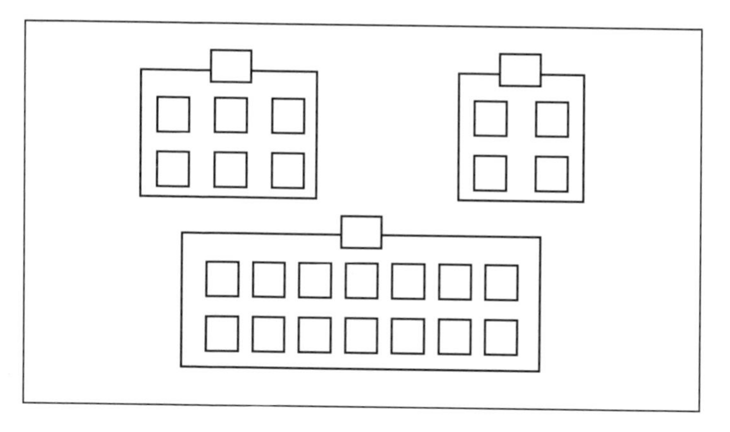

Affinity Diagram

INTERRELATIONSHIP DIGRAPH

The interrelationship digraph (ID) takes a central idea, issue, or problem and maps out the logical or sequential links among related items. It is a creative process that shows that every idea can be logically linked with more than one other idea at a time. It allows for multidirectional, rather than linear, thinking. In this program the ID is used to determine the impacts, causes, effects, and influences that exist between two items.

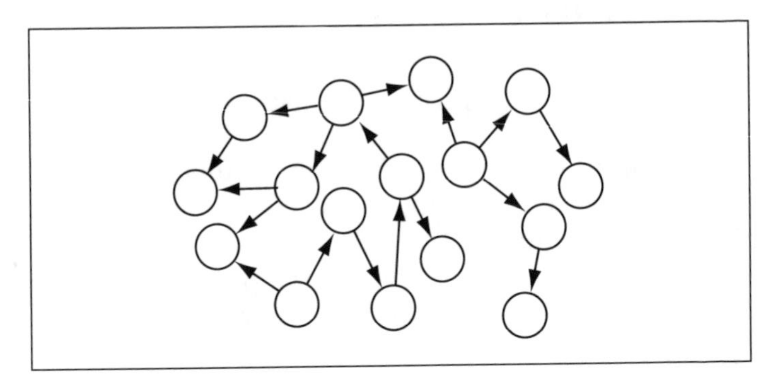

Interrelationship Digraph

TREE DIAGRAM

The tree diagram systematically maps out, in increasing detail, the full range of paths and tasks that need to be accomplished in order to achieve a primary goal and every related subgoal. In the original Japanese context, it describes the methods by which every *purpose* is to be achieved.

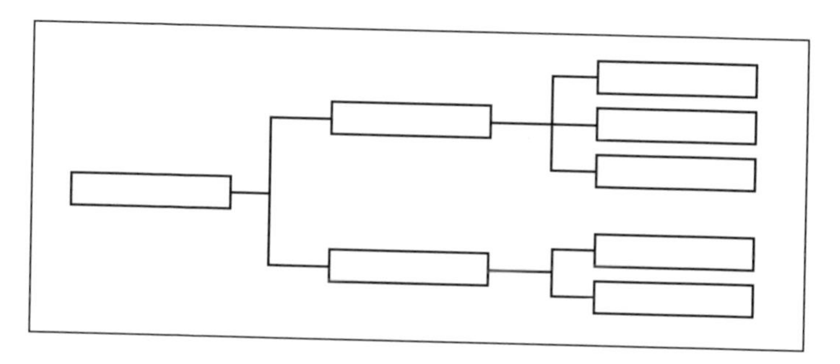

Tree Diagram

MATRIX DIAGRAM

The matrix diagram organizes large numbers of pieces of information, such as characteristics, functions, and tasks into sets of items to be compared. By graphically showing the logical connecting point between any two or more items, a matrix diagram can bring to the surface the items in each set that are related.

	a	b	c	d	e	f	g	h
1								
2								
3								
4								
5								
6								

Matrix Diagram

PRIORITIZATION MATRICES

Prioritization matrices prioritize tasks, issues, product and service characteristics, and so on, based on known weighted criteria using a combination of tree and matrix diagram techniques. Above all, they are tools for decision making.

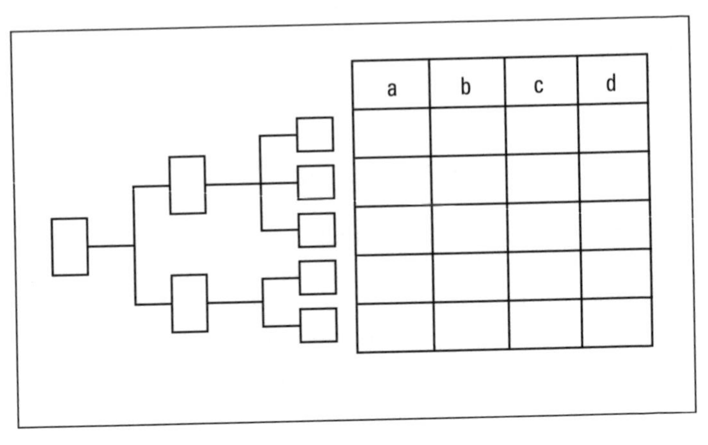

Prioritization Matrices

PROCESS DECISION PROGRAM CHART

The process decision program chart (PDPC) is a method that maps out conceivable events and contingencies that can occur in any implementation plan. It, in turn, identifies feasible countermeasures in response to these problems. This tool is used to plan each possible chain of events that needs to occur when the problem or goal is an unfamiliar one.

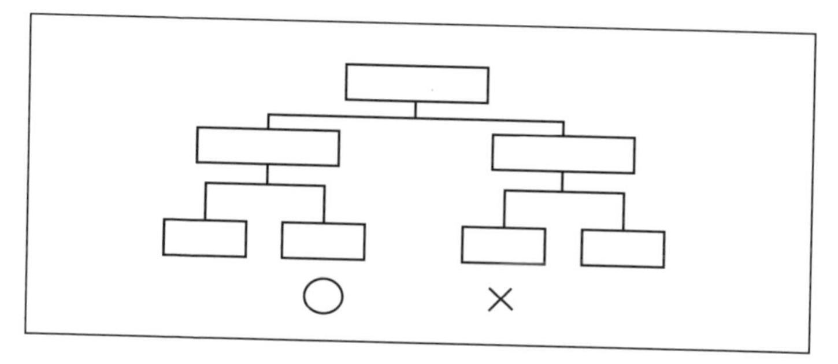

Process Decision Program Chart

ACTIVITY NETWORK DIAGRAM

The activity network diagram is used to plan the most appropriate schedule for the completion of any complex task and all of its related subtasks. It projects likely completion time and monitors all subtasks for adherence to the necessary schedule. This is used when the task at hand is a familiar one with subtasks of a known duration.

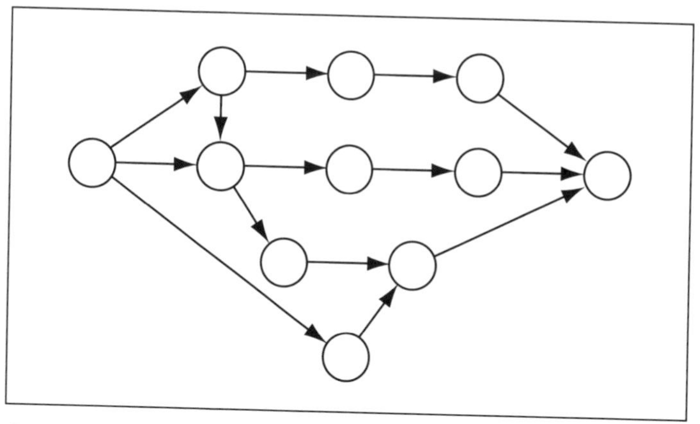

Activity Network Diagram

APPENDIX C
THE 7 CREATIVITY TOOLS

PURPOSE

The seven creativity tools are not new tools, but are presented here in a systematic way to provide widespread education and use of structured creativity in an environment of continuous improvement. A research committee at GOAL/QPC established the seven creativity tools in the mid-1990s.

PROBLEM DEFINITION

Working on the right problem is an important starting point for creativity.

The purpose hierarchy, popularized by Gerald Nadler of USC, enables one to see a whole range of problems, from the simple to the complex, and thereby pick the one that is currently the most appropriate to work on.[15] In daily management this is often the best creativity tool for problem definition.

In Quality Function Deployment or new product development, the S-curve and the lines of technological evolution, as developed by Altshuller, are best for creatively defining what to work on.[16]

The S-curve shows when product life cycle will peak and the need for a new breakthrough product. The technological evolution chart shows the natural divergence and convergence of technology.

Purpose Hierarchy	S-Curve Product Cycle	Technological Evolution
aaaaaaaa bbbbbbbb → cccccccccc dddddddddd eeeeeeeeeeeeeee ffffffffffffffffffffffffffff		

BRAINSTORMING

Brainstorming is a tool for gathering surface-level ideas from a team of people in various structured and unstructured formats. In TQM circles these ideas are often grouped by an affinity chart. It is the simplest of the tools, but is always a good starting point. If you can solve your problems here, you have done it efficiently.

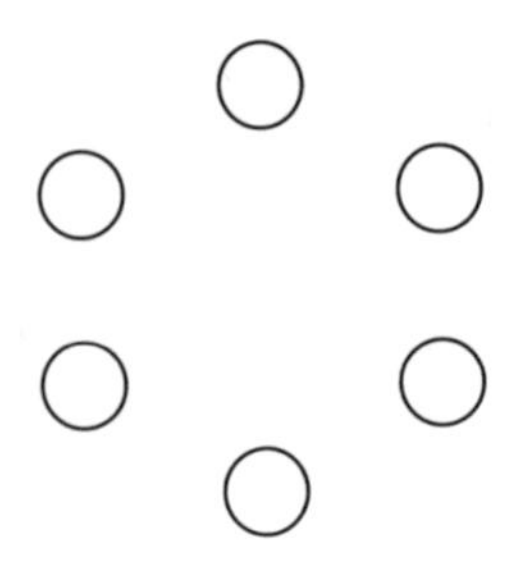

BRAINWRITING (6-5-3)

Brainwriting is a creativity tool in which ideas are written down as they occur. Sometimes this is done by taking notes to create a mind map that shows how the ideas are linked together.

Perhaps the best known form of brainwriting is the 6-5-3 technique. A group of six people each write down three ideas on a subject on the form shown below. All the forms are passed to the right. Each person then writes three additional ideas, hopefully triggered by the ideas they read from preceding person(s). This continues until the forms are finished. It is a quick way to get a lot of ideas.

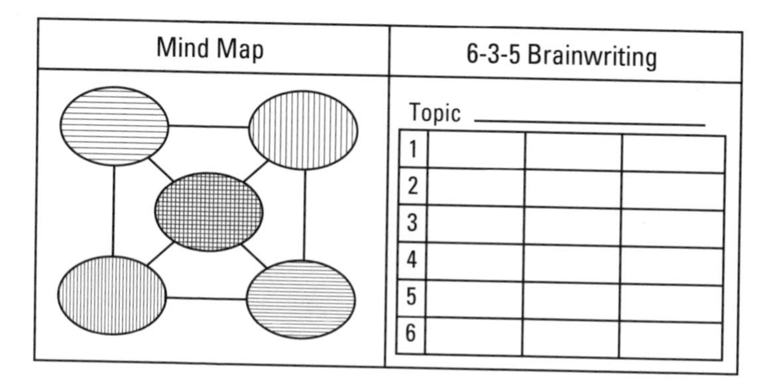

CREATIVE BRAINSTORMING

Creative brainstorming is a tool that helps one go underneath the surface ideas that easily come to consciousness by tricking the brain through a fun exercise.

First, the group brainstorms on the problem. Then they brainstorm on a silly problem and apply the ideas from the silly problem to the initial problem. Then they combine the ideas from both for the best solution.

An example would be getting more widespread use of computers. The first brainstorm would be in traditional format and generate many useful ideas. Then the second brainstorm would generate ideas for a silly problem, for example, how to get employees to wear cardboard noses. There would be a lot of laughter and crazy ideas, some of which could be very helpful for the original problem.

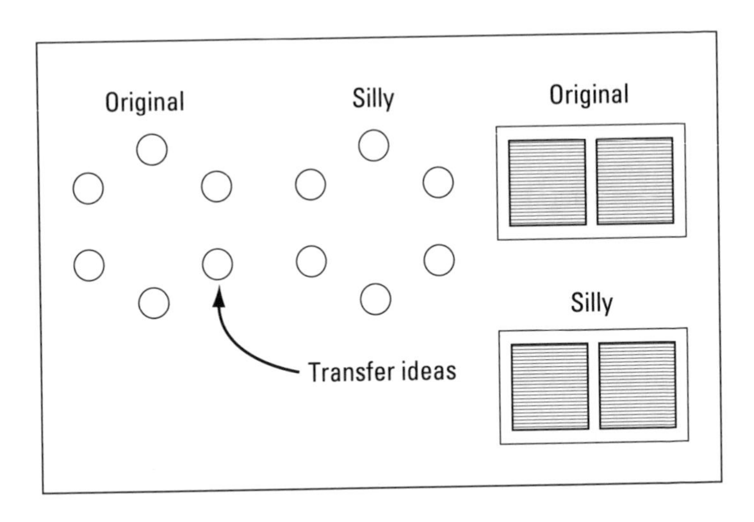

WORD AND PICTURE ASSOCIATION

Another useful tool for getting into the unconscious for creative idea generation is to use pictures or words. The pictures and words are examined to see what is the root of what is happening there and the root idea is then applied to the problem at hand.

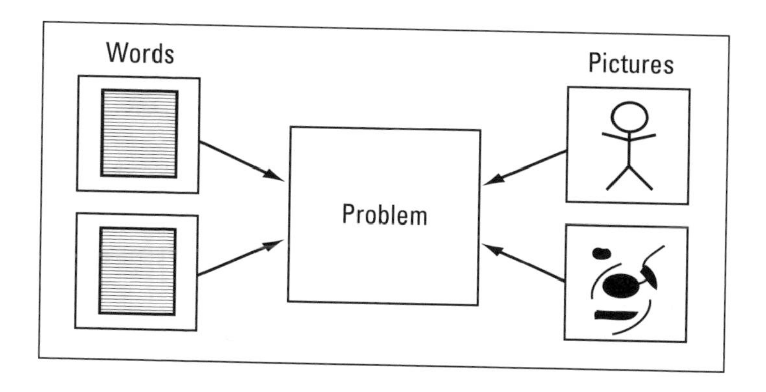

ADVANCED ANALOGIES

Advanced analogies bring association to an art form. One method, the TILMAG method, identifies the key functions or attributes of a problem and generates examples through paired comparisons. Each example is studied for its core technology or purpose, which is then applied to solving the problem. TILMAG translates as "transformation of ideal solution element in a matrix for associations and things in common."[17]

Systematic innovation (also referred to as TRIZ/TIPS) defines problems in terms of their root contradictions. Positive and negative effects are looked up on the principle matrix to find what principles might apply. Examples of the principles are then studied for their core technology, which is applied to the problem. Because Altshuller developed the principles and examples from the review of over one million patents, this tool is a very powerful creative tool for pushing someone out of his or her traditional boxes. TRIZ translates as "theory of creative problem solving," and TIPS is an acronym for "theory of inventive problem solving."[18]

TILMAG method

	1	2	3	4	5	6
7						
6						
5						
4						
3						
2						

Systematic Innovation

Contradiction matrix

	1	2	3	4	5	6
a						
b						
c						
d						
e						
f						

Examples

Example #1	Example #2

MORPHOLOGICAL CHART

The morphological chart pushes creativity the furthest because it strives to identify all possible solutions to the problem. This is accomplished by carefully identifying all the possible parameters that are unique, complete, and do not overlap, then finding similarly complete and discrete elements for each parameter.

The following morphological chart of a house is a popular example:

All Solutions	House Example
Parameters	

House Example

Parameters	Examples					
Foundation shape						
Roof structure						
2nd floor position						
Roof material	Poured asphalt	Tile	Slate	Tile	Asphalt shingles	
Siding material	Stucco	Brick	Clapboard	Aluminum siding	Clapboard	
Turret						
Window structure						

GLOSSARY

Brainwriting: The process of writing down ideas as they come to mind. Also the process of using another's written ideas to stimulate one's own written ideas.

Breakthrough: A major improvement that usually results in growth of market share.

Cost engineering: A process for determining whether a function or part is too expensive for its relative value and the process for reducing this cost without reducing the quality.

Creativity: The process of generating or capturing a new idea.

Customer: The one who uses the product or service. Also the next one in a chain of steps or actions.

Customer demands: A list of the wants of the customer translated accurately from actual words to words that are understood unequivocally throughout the producer organization.

The Customer Window: A four-box diagram for sorting customer wants into those that are 1) wanted and met well, 2) wanted and not met well, 3) not wanted and met well, and 4) not wanted and not met well.

Exciting quality: Those characteristics of a product or service that surprise and delight a customer because they respond to wants in a way the supplier conceives rather than the customer.

Expected quality: The quality customers take for granted until it is missing.

House of Quality: The name given to a matrix that matches customer wants with quality characteristics (or metrics) as a basis for prioritizing quality characteristics. One version of this matrix is topped with a triangle that examines interactions of quality characteristics with each other. This triangle has been referred to as the roof of the house of quality.

Innovation: The process of successfully putting a new idea to work.

Morphological chart: A matrix whose parameters try to incorporate all the solutions to a particular problem.

New concept selection: A matrix popularized by the late Stuart Pugh of Scotland that uses pluses (+) and minuses (−) to rate and evaluate new ideas versus the status quo of a current model.

One-dimensional quality: The kind of quality that is often reflected in a specification. It is good or bad depending on how much or how little you have of this characteristic.

Purpose hierarchy: A tool, popularized by Gerald Nadler, that groups purposes and tactics in an array, from simple to complex, to ease the process of building consensus around what problem should be worked on now.

Quality characteristics: A list of measures or metrics that provide a yardstick for measuring to what extent customer demand is met.

Quality function deployment: A process by which the wants of customers are prioritized in a series of charts and tables, resulting in a breakthrough design of product or service that delights customers and grows market share.

Reliability engineering: The process of assuring that the whole product or its parts will not fail within a certain period of time (for example, in a car this is often 100,000 miles).

Seven creativity tool boxes: A series of tools to assist companywide creativity and innovation, including problem definition, for example, purpose hierarchy; brainstorming; brainwriting, e.g., 6-3-5; creative brainstorming; associations by word, picture, and so on; analogies such as synetics or TRIZ; morphological charts and boxes.

Seven management and planning tools (also called 7 new tools): Includes the affinity chart (also called KJ after Jiro Kawakito), interrelations digraph (also called relations diagram), tree diagram, matrix diagram, prioritization matrices, process decision program chart (PDPC), and arrow diagram. Some lists include matrix data analysis instead of the prioritization matrices. See Appendix B for examples.

Seven quality control tools: A group of problem-solving tools that usually include the cause-and-effect diagram (also called fishbone or Ishikawa diagram,) the run chart, the scatter diagram, the flowchart, the Pareto chart, the histogram, and the control chart. Other lists of 7 Quality Control Tools sometimes include the check sheet and graphs. See Appendix A for examples.

Total quality management: All employees in all departments, every day, improving or maintaining quality, cost, delivery, yield, procedures, and systems to give customers the product and service that is most economical and best qualified.

TRIZ: Russian acronym for "theory of creative problem solving," an array of principles arranged by Altshuller and based on examination of one million patents, enabling individuals and teams to perform systemic innovation.

Value engineering: A process for identifying which parts, characteristics, or functions of a product or service have relatively high cost compared to their value. It may also refer to the processes used to take cost out of a product or service without reducing its quality.

Voice of the customer: Describes the wants of the customer as they are captured in various forms.

Voice of the customer table (VOCT): A table for recording the voice of the customer in various categories, such as *what, when, where, why,* and *how* the product or service will be used.

Voice of the engineer: Describes what engineers believe to be essential to the successful functioning of a product or service.

NOTES

1. Noriaki Kano, "Three arrow model," *Quality Magazine,* 14, no. 2 (JUSE, Tokyo, Japan), 39-48. Reference is from translation by GOAL/QPC from the original Japanese version

2. *Customer Window™ Process: The Voice of the Customer for TQM: Participant Manual,* ARBOR, Inc. This tool was developed by David Saunders of ARBOR.

3. Gerald Nadler and Shozo Hibino, *Breakthrough Thinking* (Rocklin, Calif.: Prima Publishing and Communications, 1990), 28-29. Also in more depth, Gerald Nadler, *The Planning and Design Approach,* (Methuen, Mass.: GOAL/QPC, 1981), 115, 135-142.

4. *QFD: Advanced QFD Application Articles,* a GOAL/QPC Research Report, Methuen, MA, 1990. *QFD: A Process for Translating Customers' Needs Into a Better Product and Profit,* a GOAL/QPC Research Report, Methuen, MA, 1989.

5. Bob King, *Better Designs in Half the Time,* (Methuen, Mass.: GOAL/QPC, 1989), 4-3.

6. Michael Brassard and Diane Ritter, *The Memory Jogger II™* (Methuen , Mass.: GOAL/QPC, 1994): 156. See also *The Memory Jogger Plus™* by Michael Brassard (Methuen Mass.: GOAL/QPC, 1989): 73, and Appendix A.

7. For more information on creativity in organizations, see Edward deBono, *Serious Creativity: Using the Power of Lateral Thinking to Create New Ideas* (New York: Harper Business, 1992), Teresa Amabile, "Social Psychology of Creativity: A Consensual Assessment of Techniques," *Journal of Personality and Social Psychology* 8 (July 19, 1994b): 573-578, and Helmut Schlicksupp, *Creativity Workshop: Idea-finding, Problem Solving, and Innovation Conference Planning and Organizing* (Wurzberg, Germany: Vogel Publishing House, 1993).

8. Bob King and Helmut Schlicksupp, *The Seven Creativity Tool Boxes* (Methuen, Mass.: GOAL/QPC, 1995).

9. Bob King, *Better Designs in Half the Time,* 10-1.

10. Bob King and Helmut Schlicksupp, *The Seven Creativity Tool Boxes.*

11. Michael Brassard and Diane Ritter, *The Memory Jogger II.*™

12. Roger Milliken, as quoted by Bob Galvin of Motorola in *Leaders of Quality,* a video by GOAL/QPC, Methuen, Mass. (1992).

13. Robert C. Camp, *Benchmarking: The Search for Industry Best Practices That Lead to Superior Performance* (Milwaukee: Quality Press/Quality Resources, 1989).

14. Gerald Nadler and Shozo Hibino, *Breakthrough Thinking,* and Gerald Nadler, *The Planning and Design Approach.*

15. G.S. Altshuller, *Creativity as an Exact Science: The Theory of the Solution of Inventive Problems.* Translated by Anthony Williams, (New York: Gordon and Breach Science Publishers, 1988).

16. Horst, Geschka, Ute von Relbnitz, and Kjetli Storvik, "Idea Generation Methods: Creative Solutions to Business and Technical Problems," *Battelle Technical Inputs to Planning*, Review No. 5, (Columbus, Ohio: Battelle Memorial Institute).

17. G.S. Altshuller, *Creativity as an Exact Science: The Theory of the Solution of Inventive Problems* (translated by Anthony Williams).

FURTHER READING

Brassard, Michael and Diane Ritter, *The Memory Jogger II™* (Methuen, Mass.: GOAL/QPC, 1994), summary of 7QC and 7MP tools.

Camp, Robert, *Benchmarking: The Search for Industry Best Practices that Lead to Superior Performance* (Milwaukee: Quality Press/Quality Resources, 1989).

Cohen, Lou, *Quality Function Deployment: How to Make QFD Work for You,* (Reading, Mass.: Addison-Wesley, 1995).

King, Robert, *Better Designs in Half the Time,* (Methuen, Mass.: GOAL/QPC, 1989).

King, Robert and Helmut Schlicksupp, *The Seven Creativity Tool Boxes* (Methuen Mass.: GOAL/QPC, 1995).

Nadler, Gerald and Shozo Hibino, *Breakthrough Thinking* (Rocklin, Calif.: Prima Publishing and Communications, 1990).

ABOUT THE AUTHOR

Bob King is the executive director of GOAL/QPC. He has directed the research of U.S. quality gurus and TQM worldwide and has led research on the integration of TQM principles with American innovation. He has established several research/application committees to investigate how advanced TQM applies to health care, education, government, financial services, and other nonmanufacturing industries and has worked with many executives of leading U.S. organizations in charting their development of TQM.

Mr. King is author of two books, *Better Designs in Half the Time* and *Hoshin Planning: The Developmental Approach*. He served as an examiner for the Malcolm Baldrige National Quality Award for 1989 and 1990, and he is the lead instructor in training others to become qualified examiners for the Massachusetts Quality Award. He is currently organizing a worldwide research journal for total quality management.

PRAISE FOR THE MANAGEMENT MASTER SERIES

"A rare information resource.... Each book is a gem; each set of six books a basic library.... Handy guides for success in the '90s and the new millennium."

Otis Wolkins
Vice President Quality Services/Marketing
Administration, GTE

"Productivity Press has provided a real service in its *Management Master Series*. These little books fill the huge gap between the 'bites' of oversimplified information found in most business magazines and the full-length books that no one has enough time to read. They have chosen very important topics in quality and found well-known authors who are willing to hold themselves within the 'one plane trip's worth' length limitation. Every serious manager should have a few of these in their reading backlog to help keep up with today's new management challenges."

C. Jackson Grayson, Jr.
Chairman, American Productivity & Quality Center

"The *Management Master Series* takes the Cliffs Notes approach to management ideas, with each monograph a tight 50 pages of remarkably meaty concepts that are defined, dissected, and contextualized for easy digestion."

Industry Week

"A concise overview of the critical success factors for today's leaders."

Quality Digest

"A wonderful collection of practical advice for managers."

Edgar R. Fiedler
Vice President and Economic Counsellor,
The Conference Board

"A great resource tool for business, government, and education."

Dr. Dennis J. Murray
President, Marist College

THE MANAGEMENT MASTER SERIES

The Management Master Series offers business managers leading-edge information on the best contemporary management practices. Written by respected authorities, each short "briefcase book" addresses a specific topic in a concise, to-the-point presentation, using both text and illustrations. These are ideal books for busy managers who want to get the whole message quickly.

Set 1. Great Management Ideas

Management Alert: Don't Reform—Transform!
Michael J. Kami
Transform your corporation: adapt faster, be more productive, perform better.

Vision, Mission, Total Quality: Leadership Tools for Turbulent Times
William F. Christopher
Build your vision and mission to achieve world class goals.

The Power of Strategic Partnering
Eberhard E. Scheuing
Take advantage of the strengths in your customer-supplier chain.

New Performance Measures
Brian H. Maskell
Measure service, quality, and flexibility with methods that address your customers' needs.

Motivating Superior Performance
Saul W. Gellerman
Use these key factors—non-monetary as well as monetary—to improve employee performance.

Doing and Rewarding: Inside a High-Performance Organization
Carl G. Thor
Design systems to reward superior performance and encourage productivity.

Set 2. Total Quality

The 16-Point Strategy for Productivity and Total Quality
William F. Christopher/Carl G. Thor
Essential points you need to know to improve the performance of your organization.

The TQM Paradigm: Key Ideas That Make It Work
Derm Barrett
Get a firm grasp of the world-changing ideas beyond the Total Quality movement.

Process Management: A Systems Approach to Total Quality
Eugene H. Melan
Learn how a business process orientation will clarify and streamline your organization's capabilities.

Practical Benchmarking for Mutual Improvement
Carl G. Thor
Discover a down-to-earth approach to benchmarking and building useful partnerships for quality.

Mistake-Proofing: Designing Errors Out
Richard B. Chase and Douglas M. Stewart
Learn how to eliminate errors and defects at the source with inexpensive *poka-yoke* devices and staff creativity.

Communicating, Training, and Developing for Quality Performance
Saul W. Gellerman
Gain quick expertise in communication and employee development basics.

Set 3. Customer Focus

Designing Products and Services That Customers Want
Robert King
Here are guidelines for designing customer-exciting products and services to meet the demands for continuous improvement and constant innovation to satisfy customers.

Creating Customers for Life
Eberhard E. Scheuing
Learn how to use quality function deployment to meet the demands for continuous improvement and constant innovation to satisfy customers.

Building Bridges to Customers
Gerald A. Michaelson
From the priceless value of a single customer to balancing priorities, Michaelson delivers a powerful guide for instituting a customer-based culture within any organization.

Delivering Customer Value: It's Everyone's Job
Karl Albrecht
This volume is dedicated to empowering people to deliver customer value and aligning a company's service systems.

Shared Expectations: Sustaining Customer Relationships
Wayne A. Little
How to create a process for sharing expectations and building lasting and profitable relationships with customers and suppliers that incorporates performance goals and measures.

Service Recovery: Fixing Broken Customers
Ron Zemke
Here are the guidelines for developing a customer-retaining service recovery system that can be a strategic asset in a company's total quality effort.

PRODUCTIVITY PRESS, Dept. BK, PO Box 13390, Portland, OR 97213-0390
Telephone: 1-800-394-6868 Fax: 1-800-394-6286

Set 4. Leadership (available November, 1995)

Leading the Way to Organization Renewal

Burt Nanus

How to build and steer a continually renewing and transforming organization by applying a vision to action strategy.

Checklist for Leaders

Gabriel Hevesi

Learn to focus day-to-day decisions and actions, leadership, communications, team building, planning, and efficiency.

Creating Leaders for Tomorrow

Karl Albrecht

How to mobilize all the intelligence of the organization to create value for customers.

Total Quality: A Framework for Leadership

D. Otis Wolkins

Consider the problems and opportunities in today's world of changing technology, global competition, and rising customer expectations in terms of the leadership role.

From Management to Leadership

Lawrence M. Miller

A visionary analysis of the qualities required of leaders in today's business: vision and values, enthusiasm for customers, teamwork, and problem-solving skills at all levels.

High Performance Leadership: Creating Value in a World of Change

Leonard R. Sayles

Examine the need for leadership involvement in work systems and operations technology to meet the increasing demands for short development cycles and technologically complex products and services.

ABOUT PRODUCTIVITY PRESS

Productivity Press exists to support the continuous improvement of American business and industry.

Since 1983, Productivity has published more than 100 books on the world's best manufacturing methods and management strategies. Many Productivity Press titles are direct source materials translated for the first time into English from industrial leaders around the world.

The impact of the Productivity publishing program on Western industry has been profound. Leading companies in virtually every industry sector use Productivity Press books for education and training. These books ride the cutting edge of today's business trends and include books on total quality management (TQM), corporate management, Just-In-Time manufacturing process improvements, total employee involvement (TEI), profit management, product design and development, total productive maintenance (TPM), and system dynamics.

To get a copy of the full-color catalog, call 800-394-6868 or fax 800-394-6286.

To view sample chapters and see the complete line of books, visit the Productivity Press online catalog on the Internet at *http://www.ppress.com/*

Productivity Press titles are distributed to the trade by National Book Network, 800-462-6420

TO ORDER: Write, phone, or fax Productivity Press, Dept. BK, P.O. Box 13390, Portland, OR 97213-0390, phone 800-394-6868, fax 800-394-6286. Send check or charge to your credit card (American Express, Visa, MasterCard accepted).

U.S. ORDERS: Add $5 shipping for first book, $2 each additional for UPS surface delivery. We offer attractive quantity discounts for bulk purchases of individual titles; call for more information.

ORDER BY E-MAIL: Order 24 hours a day from anywhere in the world. Use either address:
To order: *service@ppress.com*
To view online catalog on the Internet and/or to order:

INTERNATIONAL ORDERS: Write, phone, or fax for quote and indicate shipping method desired. For international callers, telephone number is 503-235-0600 and fax number is 503-235-0909. Prepayment in U.S. dollars must accompany your order (checks must be drawn on U.S. banks). When quote is returned with payment, your order will be shipped promptly by the method requested.

NOTE: Prices are in U.S. dollars and are subject to change without notice.

PRODUCTIVITY PRESS, Dept. BK, PO Box 13390, Portland, OR 97213-0390
Telephone: 1-800-394-6868 Fax: 1-800-394-6286